Elvis Pantaleão Ferreira

Clay extraction

Elvis Pantaleão Ferreira

Clay extraction

Environmental diagnosis in a red ceramics production centre

ScienciaScripts

Imprint

Cover image: www.ingimage.com

This book is a translation from the original published under ISBN 978-3-330-76268-8.

Publisher:
Sciencia Scripts
is a trademark of
Dodo Books Indian Ocean Ltd. and OmniScriptum S.R.L publishing group

120 High Road, East Finchley, London, N2 9ED, United Kingdom
Str. Armeneasca 28/1, office 1, Chisinau MD-2012, Republic of Moldova, Europe
Managing Directors: Ieva Konstantinova, Victoria Ursu
info@omniscriptum.com

Printed at: see last page
ISBN: 978-620-8-38645-0

CONTENTS

CHAPTER 1

INTRODUCTION

Brazil's subsoil contains important mineral deposits. Some of these reserves are considered significant when compared to the rest of the world. As described by Barreto (2001) apud Valicheski (2009), the country produces around 70 mineral substances, represented by groups of metallic, non-metallic and energy minerals. Among the non-metallic minerals, clay plays an important role in the country's economy, with an estimated 1% share of the Gross Domestic Product, corresponding to around 6 billion dollars a year.

The anthropogenic activities with the greatest impact on natural areas include agriculture, livestock farming and the exploitation of mineral resources. Agriculture and livestock farming together occupy a much larger area than mineral exploitation. Mineral exploitation, however, is much more impactful from an environmental point of view, as it results in the complete devastation of areas, due to the removal of vegetation and arable soil, often causing irreversible impacts on the environment. However, according to Meyres et al. (2000), mining activities account for only 1.2 per cent of deforestation on the planet, while agriculture and grazing account for 69 per cent.

Man's use of natural resources has not always been in accordance with their characteristics and recovery capacities. However, in recent years the environmental issue has become a business concern. Companies are coming under increasing pressure from environmental organisations, the market and civil society to value their actions in terms of preserving the environment. Therefore, environmental responsibility is an important step for companies to remain in the market.

In this sense, the role of environmental villains that companies have been playing has its raison d'être, since there are proportionally few companies that repair and worry about the damage caused to the environment by their production processes, as shown by global and national statistics (Dias 2010). And even when they do, the initiative is taken more as a response to a demand from environmental organisations than because they are taking a socially and environmentally responsible stance.

In the specific case of ceramics, it should be emphasised that the role played by these production units is undeniable and essential. However, the only way forward is for companies to adopt an environmental management system, which will enable them to have a future with less impactful development. In the state of Espírito Santo, according to a sector diagnosis carried out by SEBRAE/FINDES/IEL in 2009, there are 70 industries in the red ceramics segment. These are mainly distributed around the south and north centre of the state, associated with sector unions. The municipality of São Roque do Canaã is located in the latter centre. The production of red ceramics[1] in the municipality is a long-standing economic activity that generates jobs both locally and regionally, but also contributes to the emergence of questionable environmental problems. There are eight (08) industries in the municipality producing ceramic blocks for structural sealing and roof tiles, the latter with the highest production.

To this end, the aim of this work is to present an environmental diagnosis of clay extraction areas in a red ceramics production centre, and the textural characterisation of the soil used in the manufacture of ceramics, in the municipality of São Roque do Canaã, located in the state of Espírito Santo, Brazil.

[1] The name used in this study for the industries in the study area that manufacture ceramic blocks (structural sealing blocks and roof tiles) used in civil construction, produced from soils with high clay content and which are red in colour when fired.

CHAPTER 2

OBJECTIVES

2.1. GENERAL OBJECTIVE

The purpose of this study is to present the environmental diagnosis of clay extraction areas in a red ceramics production centre located in the municipality of São Roque do Canaã in the state of Espírito Santo, Brazil.

2.2. SPECIFIC OBJECTIVES

- Present a general characterisation of the municipality;
- Present an overview of the ceramics industry in the municipality;
- Present an environmental diagnosis of the clay extraction areas and textural characterisation of the soils;
- Draw up a table of the main environmental aspects and impacts during the implementation, extraction and decommissioning of the clay deposits;
- Draw up an Interaction Matrix of the main environmental impacts of clay extraction.

CHAPTER 3

LITERATURE REVIEW

3.1. DEFINITION OF CERAMIC PRODUCT

As described by the Brazilian Micro and Small Business Support Service - SEBRAE (2008), classifying a ceramic product takes into account the use of its products, the nature of its constituents, the textural characteristics of the biscuit (base dough), as well as other ceramic, technical and economic characteristics. Based on the raw material used, traditional (or silicate) clay-based ceramics are identified, such as structural or red ceramics, white ceramics and tiles.

In this context, according to the Portuguese Ceramic Industry Association - APICER (2006) apud Pavan (2009), ceramic materials are made from raw materials classified as natural or synthetic. The most commonly used natural materials are: clay, kaolin, quartz, feldspar, calcite, dolomite, magnesite, phyllite, talc, chromite, bauxite, graphite and zirconite. Synthetic products include, among others, alumina (aluminium oxide) in different forms (calcined, electro-fused and tabular), silicon carbonate and a wide variety of inorganic chemicals.

Red ceramic products are characterised by the red colour of their products, represented by bricks, blocks, tiles, pipes, ceiling slabs, flagstones, ornamental vases, lightweight expanded clay aggregates and others. As far as raw materials are concerned, the red ceramics sector basically uses common clay, in which the mass is a mono-component type - only clay - and can be called simple or natural. The mass is generally obtained on the basis of accumulated experience, aiming for an ideal composition of plasticity and fusibility, facilitating handling and providing mechanical resistance during firing (SEBRAE, 2008).

3.2. HISTORY OF THE CERAMICS INDUSTRY IN BRAZIL

According to SEBRAE (2008), ceramics were made in Brazil more than 2,000 years ago, even before the Portuguese "discovered" the country. These included pots, dishes and other ceramic artefacts. With regard to red ceramics, the scarce and inaccurate information refers to its use in the colonial period, based on rudimentary production techniques introduced by the Jesuits, who needed bricks to build colleges and convents.

From 1549, with the arrival of Tomé de Sousa in the country, the production of building materials was stimulated for the development of better planned and more elaborate cities. In 1575 there is evidence of the use of roof tiles in the formation of the village that would become the city of São Paulo-SP. It was from this stimulus that ceramics began to develop more intensively (SEBRAE, 2008).

The first large ceramics factory in Brazil was founded in São Paulo in 1893 by four French brothers from Marseille under the name "Estabelecimentos Sacoman Frères", later changed to "Cerâmica Sacoman S.A.", which closed down in 1956. The name of the roof tiles known as "French" or "Marseilles", which they are still called today, was due to the origin of these entrepreneurs (SEBRAE, 2008).

The red ceramics sector in Brazil produces bricks, blocks, tiles, hollow elements, slabs, flagstones, red tiles, pipes and light aggregates as its main products. The diversity of products is very high due to the demands of the consumer market, which often force a variety of dimensions that end up affecting the standardisation of products.

In the last years of the 19th and early 20th centuries, according to SEBRAE (2008), there was a process of specialisation in ceramics companies, which led to a separation between potteries (producers of bricks and tiles) and "ceramics" (producers of more sophisticated items such as shackles, pipes, tiles, crockery, pots, carvings, among others). However, despite the fact that the companies located in the study area only produce ceramic blocks for structural sealing and roof tiles, they are registered as ceramics companies.

3.3. THE CERAMICS INDUSTRY IN ESPÍRITO SANTO

In Espírito Santo, there are two unions representing the sector and 55 (fifty-five) ceramics industries spread across its 78 (seventy-eight) municipalities. According to Pavan (2009), most of these companies are family-owned. However, according to a sector diagnosis carried out by SEBRAE/FINDES/IEL in 2009, there are 70 industries in the red ceramics segment.The gross extraction of clays in Espírito Santo is 147,385 tonnes for common clays and 89,120 tonnes for refractory clays, DNPM - ANUÁRIO MINERAL BRASILEIRO - 2006. Based on this data, the state's gross production accounts for 0.34% of Brazilian production (Pavan, 2009). According to the SINDIOLARIA NORTE - Pottery Industry Union of the Centre-North Region of the State of Espírito Santo and surveys carried out, 41 companies are located in 11 municipalities. There are two industrial centres in this region: the north (Boa Esperança and Nova Venécia) and the centre (Colatina, Baixo Guandu, Governador Lindenberg, Ibiraçu, João Neiva, Linhares, Marilândia, Santa Teresa and São Roque do Canaã).

The northern industrial cluster is made up of the municipalities of Boa Esperança and Nova Venécia, as shown in Table 1, where there are three unionised companies and two non-unionised ones, for a total of five ceramics industries.

Table 1 - Cities that make up the northern industrial estate

City	Unionised	Non-unionised
Good Hope	2	**1**
Nova Venécia	1	**1**
Total	**3**	**2**

SOURCE: SINDIOLARIA NORTE, 2011.

The central industrial hub is made up of 9 (nine) municipalities and, as shown in Table 2, has 30 (thirty) unionised companies and 6 (six) non-unionised ones, totalling 36 (thirty-six) ceramics industries.

Table 2 - Cities that make up the industrial centre

City	Unionised	Non-unionised
Baixo Guandu	2	-
Colatina	6	2
Governor Lindenberg	3	-
Ibiraçu	1	-
João Neiva	2	-
Linhares	4	3
Marilance	2	-
Santa Teresa	2	-
São Roque do Canaã	8	1

Total	30	6

SOURCE: SINDIOLARIA NORTE, 2011.

According to the SINDIOLARIA SUL - Pottery Industry Union of the Southern Region of the State of Espírito Santo and surveys carried out, 14 companies are located in 7 municipalities, 9 of which are unionised and 5 of which are non-unionised, forming the southern industrial cluster. These figures are shown in Table 3.

Table 3 - Cities that make up the industrial centre of the southern region

City	Unionised	Non-unionised
Anchieta	0	**1**
Cachoeiro do Itapemirim	2	**2**
Castle	1	**0**
Iconha	0	**0**
Itapemirim	5	**1**
Piúma	0	**1**
New River	1	**0**
Total	**9**	**5**

SOURCE: SINDIOLARIA NORTE, 2011.

3.4. THE ENVIRONMENTAL MANAGEMENT SYSTEM IN COMPANIES

According to Valverde (2005), environmental management can be defined as a system that includes organisational structure, planning activities,

responsibilities, practices, procedures, processes and resources to develop, implement, achieve, critically analyse and maintain environmental policy. Thus, environmental management is the way in which the organisation mobilises itself, internally and externally, to achieve the desired environmental quality.

From a business point of view, Dias (2010) mentions that environmental management is the expression used to describe business management that is geared towards avoiding problems for the environment as far as possible. In other words, it is management whose aim is to ensure that environmental effects do not exceed the carrying capacity of the environment in which the organisation is located, i.e. to achieve sustainable development.

Corporate environmental management is restricted to companies and institutions and can be defined as a set of administrative and operational policies, programmes and practices that take into account the health and safety of people and the protection of the environment by eliminating or reducing environmental impacts and damage resulting from the planning, implementation, operation, expansion, relocation or decommissioning of undertakings or activities, completing all phases of the product's life cycle (QUEZADA; PIERRE, 1998 apud VALVERDE, 2005).

Dias (2010) emphasises that environmental management is the main instrument for achieving sustainable industrial development. The process of environmental management in companies is deeply linked to the rules that are drawn up by public institutions (town halls, state and federal governments) on the environment.

Still according to the author, environmental management is above all a concept, a conception of how the use of natural resources should be managed, through economic actions or measures, investments and institutional and legal measures, with the aim of maintaining or recovering the quality of the environment and ensuring the productivity of resources and social development.

According to Valverde (2005), analysing and implementing environmental management mechanisms in companies, as well as using means, instruments and models, such as the citizen company, can meet social concerns with regard to environmental issues.

The Environmental Management System, along these lines, can be conceptualised as the part of the overall management system that includes the organisational structure, planning activities, responsibilities, practices, procedures, processes and resources for developing, implementing, achieving, critically analysing and maintaining environmental policy (ABNT apud VALVERDE, 2005).

According to DIAS (2010), with a view to implementing management processes, methods and certification systems have been devised that aim to "unify language and actions" in the company. The ISO 14.000 standards are examples of this model, characterised as voluntary standards, and present guidelines for the development and implementation of environmental management systems.

Although there isn't much discussion about environmental management in companies without recourse to the ISO 14,000 standards, environmental management systems cannot be thought of solely in terms of the pursuit of such certification or the implementation of the mechanisms it enshrines. This would actually diminish the importance of environmental management. There needs to be the development of a greater awareness that seeks to internalise, making its participants true citizens (DIAS, 2010).

3.5. COMPANIES AND THE ENVIRONMENT

Companies are mainly responsible for the depletion and alterations to natural resources, from which they obtain raw materials to make products that will be used by people. According to Dias (2010), although this activity is of great use, in recent years it has almost taken a back seat due to the environmental problems caused by industries, these problems becoming the most visible aspect, most of the time, of their relationship with the natural environment.

However, the role of environmental villains that companies have been playing is justified, as there are proportionally few companies that are concerned and make their production processes more ecologically efficient, as shown by global and national statistics (SÁNCHEZ 2008). And even when they do, the initiative is taken more as a response to a demand from government bodies than because they are taking a socially and environmentally responsible stance.

According to DIAS (2010), there are a number of external factors that cause companies to respond in order to reduce or avoid environmental changes and contamination. These include the state, the local community and the consumer market. The environmental legislation applied by all levels of state, by the environmental control and regulation institutions, limits the company's freedom to interfere in the environment. The state uses these legal instruments to protect the environment and social welfare (SÁNCHEZ 2008).

The role of the state is characterised by regulation, which can be classified into two main groups: command and control and the adoption of economic instruments.

Under the first, more traditional method, the government establishes regulations for the use of environmental resources and begins to monitor compliance with the legislation, punishing any offenders, thus basing itself on the normative pressure of established standards (DIAS, 2010).

According to the author, with the use of economic instruments, the second method, the prices of environmental goods should reflect, as accurately as possible, the values attributed to them by society, so that the use of these goods can be charged for, either directly or indirectly.

The local communities where business units are located, in turn, also act in the form of informal regulation, as they are increasingly becoming important players in relation to contamination problems, as they are the first to suffer the consequences of pollution or alterations to the environment, and as a result have a quicker response capacity, affecting companies' decisions regarding greater environmental control (SÁNCHEZ 2008).Finally, there is the role of the market as a regulatory agent, where there is a growing increase in environmental awareness, which according to Dias (2010), varies according to each market. The company is forced to adopt actions to obtain raw materials that avoid environmental liabilities, for which its consumer market demands an environmentally correct production chain. The most developed countries, the most developed regions of a country are the ones that consume the most environmentally friendly products, and this involves the company's reputation as a benefactor or not of the environment.

3.6. COMPANIES AND THE LOCAL COMMUNITY

Worsening environmental conditions have led to an increase in public awareness of the importance of the natural environment. In this sense, Dias (2010) mentions that societies are increasing their demands on the agents most directly involved, particularly public administrations and companies. In the case of public authorities, they are responsible for the common good, and in the case of companies, they are the main visible agents of environmental change.

Under normal circumstances, the population receives a large amount of information about environmental problems through the media (the ozone layer, the greenhouse effect, the disappearance of species and excessive pollution are among the most frequently discussed), thus developing, at first, a form of conservationist awareness, which then quickly turns into protectionist awareness (Dias, 2010). At the same time, there is a growing realisation that protecting nature makes the environment healthy and consequently improves the quality of life.

Finally, it is important to stress that the local community should be aware of the company's environmental policy, philosophy, guidelines, targets and impact minimisation programme. In this way, they will be able to assess the real and potential impact on their quality of life. Communication, on the other hand, makes the company aware of the community's wishes and level of support. A good relationship between the company and the community not only brings benefits to the latter, but also reduces the company's environmental vulnerability and increases security in the viability of the enterprise.

3.7. ENVIRONMENTAL DIAGNOSIS

The noun diagnosis from the Greek "diagnostikós" means the knowledge or determination of a disease by its symptoms or the set of data on which this determination is based. Hence, environmental diagnosis can be defined as knowledge of all the environmental components of a given area (country, state, river basin, municipality) in order to characterise its environmental quality (REDE AMBIENTE, 2011).

Diagnosis is an element in the administration of any geographical area. Environmental diagnosis is the study of the agents that cause environmental degradation in a given area, their levels of pollution, as well as the environmental conditioning factors that aggravate or reduce the effects they have on the environment. It is the interpretation of the quality situation of an environmental system or area, based on the study of the interactions and dynamics of its components, whether related to physical and biological elements or to socio-cultural factors (FEEMA, 1997).

The characterisation of the environmental situation or quality can be carried out with different objectives. One of them is, as planning methodologies advocate, to serve as a basis for knowledge and examination of the environmental situation, with a view to drawing up lines of action or decision-making to prevent, control and correct environmental problems (environmental policies and environmental management programmes).

According to Sánchez (2008), another use of the term environmental diagnosis that has become widespread in Brazil is in relation to the environmental studies required for licensing. Various types of environmental studies have been created with the aim of providing information and technical analyses to support the licensing process, as can be seen in CONAMA Resolution No. 237 of 19 December 1997 (Art. 1, item III), which defines environmental studies as:

> "These are any and all studies relating to the environmental aspects related to the location, installation, operation and expansion of an activity or undertaking, presented as a subsidy for the analysis of the licence required, such as: environmental report, environmental control plan and project, preliminary environmental report, environmental diagnosis, management plan, degraded area recovery plan and preliminary risk analysis."

Generally speaking, the aim of the Environmental Diagnosis is to present the main elements of the physical, biotic and socio-economic environment (description of geology, pedology, geomorphology, meteorology, water quality, air quality, terrestrial and aquatic fauna and flora, endangered species, studies on population dynamics, standard of living, productive structure and services) that could be modified by the implementation and operation of the project (AGENDA VERDE, 2006; REDE AMBIENTE, 2011).

3.7.1. Identification of Environmental Aspects and Impacts

The ISO 14001 standard (2004) introduced the term "environmental aspect".

This term was unknown to professionals involved in environmental impact assessment, or was used with another connotation (SÁNCHEZ, 2008). However, due to the ISO 14,000 series, it has slowly been incorporated into the vocabulary of industry professionals and consultants, and has also reached government bodies.

According to the aforementioned standard, an environmental aspect is defined as "an element of an organisation's activities, products or services that can interact with the environment". This definition, according to Sánchez (2008), requires examples; typical situations could be the emission of pollutants, the generation of waste, noise or vibrations. Other environmental aspects are those linked to the consumption of natural resources. Consuming water (a renewable resource) reduces its availability for other uses or for its ecological functions.

According to Barbieri (2006) *apud* Heuser (2007), "The identification of environmental aspects is an ongoing process that determines the impact, positive or negative, past, present and potential, of the organisation's activities on the environment."

Another concept defined by the 2004 NBR ISO 14.001 standard is the definition of environmental impact as "any adverse or beneficial modification of the environment that results, in whole or in part, from the environmental aspects of the organisation". For a better understanding of the concept, the definition of the environment should be presented as "the surroundings in which an organisation operates, including air, water, soil, natural resources, flora, fauna, human beings and their interrelationships" Heuser (2007).

According to Sánchez (2008), another definition of environmental impact is that contained in CONAMA Resolution No. 1/86, Article 1:

> **"Any alteration of** the physical, chemical or biological **properties of** the environment, caused by any form of matter or energy resulting from human activities, which directly or indirectly affects it:
> I - the health, safety and well-being of the population;
> II - social and economic activities;
> III- the aesthetic and sanitary conditions of the environment;
> **IV - the quality of environmental resources".**

Sánchez (2008) emphasises that this definition is inappropriate; it is in fact a definition of environmental pollution. Environmental impact is a broader concept and substantially distinct from pollution, while pollution has only a negative connotation, environmental impact can be beneficial or adverse (positive or negative). Paradoxically, the definition of pollution given by the National Environmental Policy Law better reflects the concept of environmental impact, although only in terms of negative impacts.

According to the author, human actions on the environment can be beneficial or adverse (positive or negative), depending on the quality of the intervention. If used correctly, they can make a huge contribution to ensuring that the human impact on nature is positive rather than negative, and depending on the type of change, it can be ecological, social and/or economic.

According to Sánchez (2008), environmental impact is the "alteration of environmental quality that results from the modification of natural or social processes caused by human action". He adds that environmental impact is the result of human action, which is its cause. Therefore, the cause should not be confused with the consequence. For example, a motorway is not an environmental impact, it causes environmental impacts.

There are various ways of identifying, analysing or presenting impacts. Some of the tools used to facilitate understanding are checklists, interaction networks, interaction diagrams and different types of matrices. This study used the interaction matrix tool.

Although the name "matrix" suggests a mathematical operation, impact identification matrices only have this name because of their shape. In fact, a matrix is made up of two lists, arranged in rows and columns. Each row lists the main activities or actions that make up the project being analysed and the other row lists the main components or elements of the environmental system. The aim is to identify the possible interactions between the components of the project or enterprise and the elements of the environment (SÁNCHEZ, 2008).

One of the first tools in matrix format dates back to 1971 and results from the work of Leopold et al. (1971), from the United States Geological Survey. Leopold and his team were pioneers in the way of systematising impact analyses, presenting a list of one hundred (100) human actions that can cause environmental impacts, and another list of eighty-eight (88) environmental components that can be affected by human actions.

Today there are countless variations of the Leopold matrix, which actually have little to do with the original. Except for the way the rows and columns are presented and organised. However, they all have the same purpose, as described by Leopold et al., (1971) apud Sánchez (2008), they have the function of communication, that is, the purpose of "summarising the environmental assessment text", thus enabling "various readers of the impact studies to quickly determine which impacts are considered significant and their relative importance.

3.7.2. Textural characterisation of the soil

The matrix or skeleton of the soil is made up of mineral solids originating from the parent rock and organic solids, and can be described using particle size analysis. Particles are stones, pebbles, gravel, sand, silt and clay. Particles smaller than 2 mm in diameter (sand, silt and clay) are the most important, as many of the physical and chemical properties of the mineral portion of the soil depend on the proportion it contains of these small particles. Therefore, only the sand, silt and clay fractions are usually considered to designate the textural class to which the analysed soil belongs (BERTONI & NETO 2008; KLEIN 2008).

According to Ferreira and Júnior (2001), soil texture can be determined using two methods. A field test, where the sensitivity to touch is correlated with the size and distribution of the soil's unit particles. This assessment is highly subjective and therefore prone to error. Texture determination in the laboratory, where the aim is to quantify the distribution of particles smaller than 2 mm. According to the author, the success of textural analysis depends on obtaining soil suspensions in which the particles are truly individualised and remain so until they are separated and quantified.

Various methods (analytical methods) have been used for textural analysis in the laboratory, but the pipette method is the one that gives the most accurate results and is therefore recommended by the Brazilian Agricultural Research Corporation (EMBRAPA). It is a sedimentation method based on the speed at which the particles that make up the soil fall. Using a pipette to collect an aliquot at the depth and time determined for the determination of clay, the sand fraction is separated by sieving and the silt corresponds to the complement of the other fractions to 100% of the original sample (EMBRAPA, 1997).

Once the proportions of sand, silt and clay are known through textural analysis, the soil's textural classification is determined. To do this, diagrams or textural triangles are used. There are various models of textural triangles. In Brazil, two triangles are used: the Brazilian Society of Soil Science (SBSC), based on the classification system proposed by the United States Department of Agriculture (USDA), and the Agronomic Institute of Campinas (AIC), based on the classification system of the International Society of Soil Science (ISSS) (FERREIRA; JÚNIOR 2001). The model used for the textural classification of the soil at the quarries in this study was that used by EMBRAPA, based on the classification system of the Brazilian Society of Soil Science - SBSC.

The clay fraction is identified as the set of particles with an equivalent diameter of less than 0.002 mm; it has the following characteristics: it is plastic and sticky when wet, hard and very cohesive when dry, has high hygroscopicity, high specific surface area, high cation exchange capacity3 - CEC, very small pores, contraction and expansion, and forms aggregates with other particles (FERREIRA; JÚNIOR 2001).

According to the author, the silt fraction is made up of particles with an equivalent diameter of between 0.002 and 0.05 mm; it has the following characteristics: silky to the touch, slightly cohesive when dry, intermediate pore size, slight or low hygroscopicity, intermediate specific surface area and low ion exchange capacity. Finally, the sand fraction is made up of particles with an equivalent diameter of between 0.05 and 2.0 mm and has the following characteristics: being loose, with grains that do not form aggregates, non-plastic, non-sticky, non-hygroscopic, predominantly large pores, small specific surface area and practically absent cation exchange capacity.

Finally, soil texture is one of the most stable physical characteristics because it is inherent to the soil and is related to the characteristics of the original material and the natural formation agents. Its great stability means that the granulometric composition of fractions smaller than 2 mm is considered an intrinsic characteristic of the material itself and is little altered by external variables. For this reason, texture is considered to be of great importance in describing, identifying and classifying soil, and to a large extent determines the economic value of the area (KLEIN 2008).

3.7.3. Permanent Preservation Areas - APP

According to the Brazilian Forest Code, Permanent Preservation Areas (APP) are areas "... covered or not by native vegetation, with the environmental function of preserving water resources, the landscape, geological stability, biodiversity, the gene flow of fauna and flora, protecting the soil and ensuring the well-being of the human population". They differ from the "Legal Reserve" areas, also defined in the same Code, in that they are not subject to exploitation of any kind, as can occur in the case of the Legal Reserve.

All the changes arising from the adoption of inadequate practices associated with the maintenance of these areas go beyond the boundaries of a rural unit, acquiring, as a whole, great social importance with impacts on the urban environment, affecting the whole of society. One of the most emblematic examples of this is the question of the availability of water resources, where the frequent shortage of water for supply in various urban centres, as well as the rationing of electricity supplies caused by the low level of reservoirs, could be attributed, in part, to the chronic degradation of riparian forests and wetland areas in various Brazilian river basins in recent decades (FARIAS et. al, 2006).

PPAs are of fundamental importance to the environment, since the benefits provided by these systems are directly related to the ecological services provided by the existing flora, including all the associations it provides with the biotic and symbiotic components of the environment (FARIAS et. al, 2006).

According to the authors, some examples of the importance of PPAs can be described as follows: in agricultural areas, it prevents or stabilises erosion processes; in spring areas, the vegetation acts as a buffer against rainfall, preventing its direct impact on the soil and its gradual compaction. Together with the mass of plant roots, it allows the soil to remain porous and able to absorb rainwater, feeding the water table; on the banks of watercourses or reservoirs, it ensures the stabilisation of their banks by preventing soil from being carried directly into the riverbed, acting as a filter.

This interface between agricultural and grazing areas and the aquatic environment allows them to participate in controlling soil erosion and water quality, avoiding the direct transport of sediments, nutrients and chemicals from higher parts of the land into the aquatic environment; in the hydrological control of a river basin, regulating the flow of surface and sub-surface water, and thus the water table; refuge and food for terrestrial and aquatic fauna; providing refuge and food for insects that pollinate crops, among others.

CHAPTER 4

METHODOLOGY

The study to characterise the proposed objectives was carried out by collecting primary and secondary data. The primary sources were acquired through on-site visits and interviews with representatives of the ceramics sector, workers and the local population. Secondary data was based on bibliographies on environmental legislation and ceramics in Brazil and the state of Espírito Santo.

The physical analyses for the textural characterisation of the soils were carried out at the Soil Chemistry and Physics Laboratory of the Federal Institute of Espírito Santo - IFES Santa Teresa campus, according to the methodology of EMBRAPA (1997). The deformed soil samples were air-dried and passed through a 2 mm sieve to obtain the fine air-dried soil (TFSA). The granulometry was carried out using the pipette method, using sodium hydroxide (NaOH Imol.L^{-1}) as a dispersing agent and slow agitation (50 rpm) for 12 hours in a Wagner-type agitator. The sand was quantified by wet sieving with a 0.053 mm mesh sieve, the clay by sedimentation, quantified by removing a 25 ml aliquot with the pipette, and the silt by the difference between the fractions.

4.1 DELIMITATION OF THE STUDY AREA

The study area is located in the Central Mesoregion of Espírito Santo in the municipality of São Roque do Canaã, with 19° 44' 23" S and 40° 39' 24" W as the geographical coordinates of the town centre. It has a territorial area of 342,395 Km^2 and a population of 11,273 inhabitants, bordered to the north by the municipality of Colatina, to the south by Santa Teresa, to the east by João Neiva and to the west by Itaguaçu. São Roque do Canaã is approximately 120 km from Vitória, the state capital, and the main access to the municipality is via the ES 080 motorway (IBGE, 2010).

In the state of Espírito Santo there are 70 (seventy) industries in the red ceramics sector. The municipality of São Roque do Canaã is part of the northern cluster, with eight industries registered with the Pottery Industry Union of the North Central Region, producing mainly ceramic blocks for structural sealing and roof tiles (SEBRE/FINDES/IEL, 2009). The study comprised the eight industries in the ceramics segment.

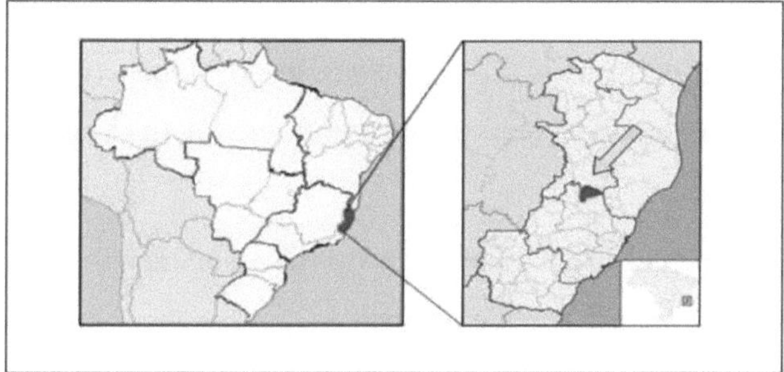

Figure 1- Location of the municipality of São Roque do Canaã - ES

Source: the author, 2011.

4.2. GENERAL CHARACTERISATION OF THE MUNICIPALITY

4.2.1 PHYSICAL ENVIRONMENT

4.2.1.1 Hydrography

The municipality is located in the Santa Maria do Doce River Basin, with an area of approximately 995.3 km^2 (nine hundred and ninety-five square kilometres). It covers all or part of the municipalities of Santa Maria de Jetibá, Santa Teresa, São Roque do Canaã and Colatina. Its main tributaries on the right bank are the Serra dos Pregos stream, the Caldeirão stream, the Santo Hilário stream, the Cinco de Novembro stream, the Vinte e Cinco de Julho stream and the Mutum stream. On the left bank, the main contributors are the Córrego do Gelo, Córrego da Onça, Rio Perdido, Rio Santa Júlia and Córrego Senador.

The Santa Maria do Doce River rises at an altitude of 1,000 metres in the Serra do Gelo, in the municipality of Santa Teresa, and runs for 85 km from its main source to its mouth on the Doce River in the city of Colatina (HYDROGRAPHIC BASIN COMMITTEE, 2011). The municipality is also part of three micro-basins, which make up the Santa Maria do Doce river basin, which contains all or part of its tributaries:

a) Santa Júlia River Sub-Basin - with the following main tributaries: Córrego Jacutinga, Córrego Palmital, Córrego Misterioso, Córrego Seco, Córrego Tancredo (Alto Tancredo and Córrego Tancredinho) among others.
b) Sub-basin of the Mutum or Boapaba River - whose main tributaries are: Córrego São Jacinto, Córrego Cachoeira do Mutum, Córrego Picadão do Mutum (on the border with Colatina) and Córrego São Miguel (municipality of Colatina) among others.
c) Triunfo River Sub-basin - whose main tributaries are the Bonsucesso Stream, which rises in the municipality of Santa Teresa. Among others.

3.7.3.1. Soil

The soils found in the Rio Doce catchment area, which includes the municipality of São Roque do Canaã, include Red-Yellow Latosols, Red and Yellow Argisols, as well as Cambisols and Gleissolos Háplicos. The predominant soils are the Red-Yellow Latosols, which are deep, steeply drained and have low natural fertility (PARH Santa Maria do Doce, 2010).

The soils in the municipality of São Roque do Canaã have been characterised as Eutrophic Red-Yellow Latosols (75%), Dystrophic Latosols (15%) and Lithosols (around 10%). In the lower altitude regions, there are also patches of Fuvic Neosols and Yellow and Red Argisols (FEITOZA, 1986; PROATER, 2008).

3.7.3.2. Weather

According to the Koppen climate classification, the climate of the São Roque do Canaã municipality is of the Cwa type, subtropical mesothermal, with a dry season in winter and a hot summer with heavy rainfall. The average annual temperature is 23°C and the average annual rainfall is 1003 mm (PERRONE; MOREIRA, 2005, PROATER, 2008).

The municipality has three natural climate zones, distributed as follows: Zone 1, with 2.6% of the territory, is characterised by cold, hilly and rainy land; Zone 2, with 13.4% of the territory, has mild, hilly and rainy/dry land; and Zone 3, with 84% of the territory, has hilly and dry land, as shown in the figure below (EMCAPA/NEPUT, 1999).

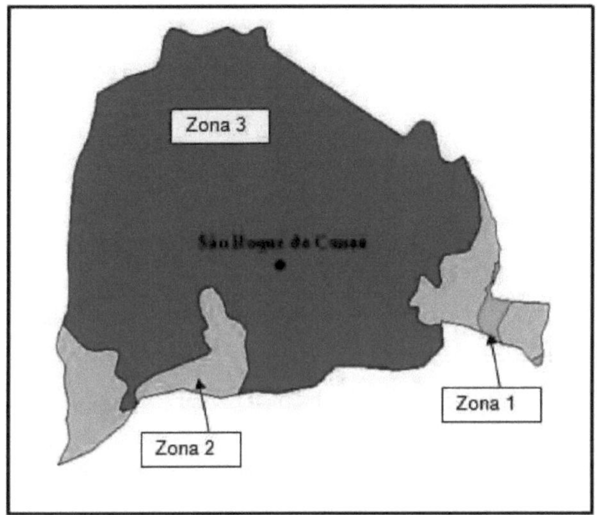

Figure 2- Climatic characterisation of the municipality

Table 1 - Characteristics of the municipality's areas

		Area (%)
Zone 1	Cold, hilly and rainy terrain	2,6
Zone 2	Mild, hilly and rainy - dry lands	13,4
Zone 3	Hot, hilly and dry land	84,0

Source: (EMCAPA/NEPUT, 1999).

Table 2 - Rainfall characteristics of the municipality

Zones	Months JFMAMJ JASOND											
Zone 1	U	U	U	U	P	P	P	S	U	U	U	U
Zone 2	U	U	U	U	P	S	S	S	S	U	U	U
Zone 3	U	P	P	P	S	S	S	S	S	U	U	U

U - rainy, S - dry, P - partially dry. Source: (EMCAPA/NEPUT, 1999).

Table 3 - Historical rainfall averages in the municipality

Jan	Feb	Sea	Apr	May	June	Jul	Aug	Set	Out	Nov	Ten	**Year**
142	**102**	**104**	**51**	**28**	**22**	**26**	**11**	**32**	**88**	**180**	**218**	**1.003**

Source: Technical Bulletin No. 7 - EMCAPA - 1981.

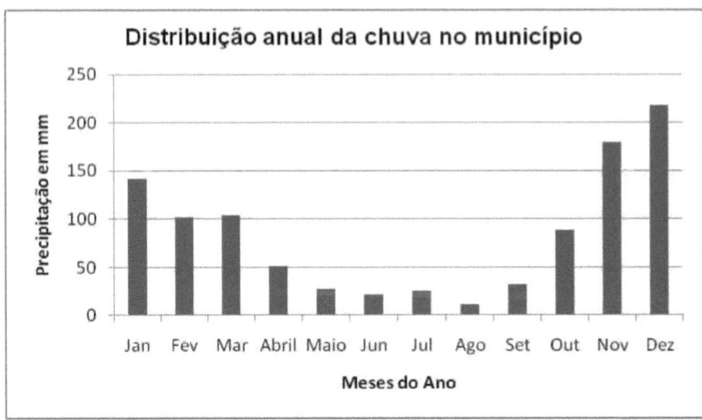

Table 4 - Distribution of rainfall according to the months of the year

Source: PROATER, modified (2008).

4.2.1.4. Relief

The relief of the region is not homogeneous, with parts of the topography being flat, slightly undulating and others with a steep slope, with the presence of rocky outcrops on the surface.Geomorphologically, the predominant topography of the municipality is undulating, followed by mountainous, flat and sloping, as shown in the table below:

Table 4 - Relief characterisation

Soil slope (%)	Area - km^2	% of the municipality's area
Up to 8% - Plan	68, 479	20
From 8 to 45 % - Wavy	136, 958	40
From 45 to 75 % - Mountainous	102, 716	30
Above 75 % - Steep	34, 242	10
Total	**342, 395**	**100**

SOURCE: PROATER, 2008.

In areas with steep slopes, problems of slope instability are common, with the occurrence of furrows and gullies causing landslides, landslides and mudflows mobilising fractured blocks. These processes contribute to the production of sediment that is discharged into water bodies. In turn, the municipality is located in a region where more than 85% of the area is highly to moderately susceptible to erosion (PARH Santa Maria do Doce, 2010).

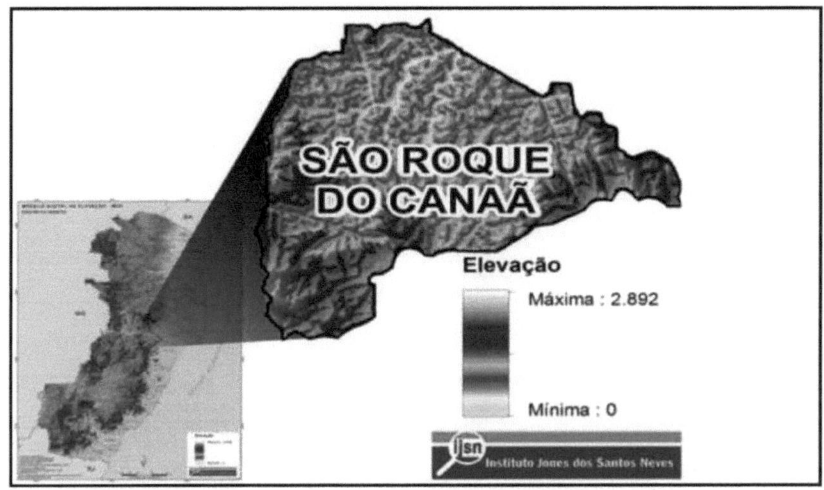

Figure 3 - Characterisation of the Municipality's Relief

Source: Jones dos Santos Neves Institute, 2011.

4.2.2 BIOLOGICAL ENVIRONMENT

4.2.2.1. Flora

According to the Brazilian Institute of Geography and Statistics (Instituto Brasileiro de Geografia e Estatística - IBGE), the municipality is under the control of the Atlantic Forest biome. However, there is a vast area of deforestation, with around 3% of its total area remaining native forest (PROATER, 2008; IBGE, 2010). Agricultural activity is predominant in the region, mainly pasture and coffee cultivation, where it is generally located up to the middle of the slopes, as is the case with pasture. In the more rugged parts, such as the mountainous edge of the plateau, anthropogenic activity is less intense, as the terrain restricts occupation, and it is in these areas that the municipality's forest remnants are concentrated.

Pasture and coffee plantations are the most prominent vegetation cover in the region, generally located up to the middle of the slopes. In the more rugged parts, such as the mountainous edge of the plateaus, anthropogenic action is less intense, as the relief restricts occupation and it is in these areas that the forest remnants are concentrated (figure 4).

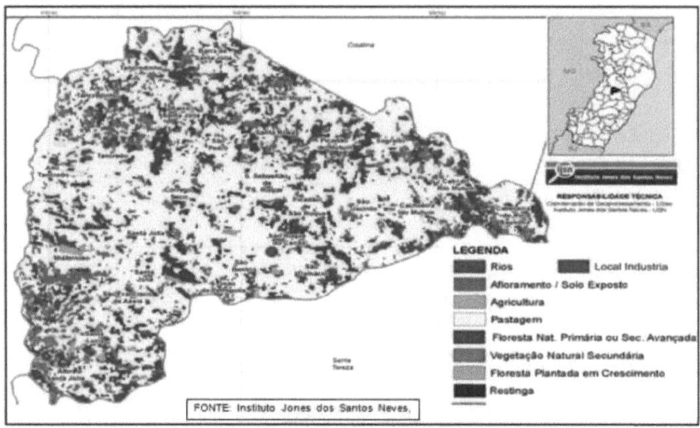

Figure 4 - Land Use and Occupation in the Municipality

4.2.2.2. Fauna

As there is no literature on the local fauna, it is worth mentioning that the municipality is under the domain of the Atlantic Forest biome, so its local fauna also characterises those that predominate in that biome. According to the Brazilian Institute of Geography and Statistics - IBGE, the state of Espírito Santo has the following families of endangered fauna:

Table 5 - Endangered bird families

Fauna	Family	Family
Birds	Accipitridae	Motacillidae
	Anatity	Muscicapidae
	Ardeidae	Phaethontidae
	Caprimulgidae	Phasianidae
	Conopophagidae	Picidae
	Cotingidae	Piridae
	Cracidae	Procellariidae
	Cuculidae	Psittacidae
	Dendrocolaptidae	Psophiliidae

Diomedeidae	Rallidae
Embrerizidae	Ramphastidae
Formicariidae	Rhinocryptidae
Fregatidae	Thamnophlidae
Fringillidae	Trochilidae
Laridae	Tyrannidae
Momotidae	Vireonidae

SOURCE: IBGE Interactive Maps, 2005.

Table 6 - Endangered amphibian families

Fauna	Family
Amphibia ns	Leoptodactylidae

Source: IBGE Mapas Interativos, 2005.

Table 7 - Reptile families threatened with extinction

Fauna	Family
	Chelidae
	Cheloniidae
Reptiles	Crocodilidae
	Dermochhelyidae
	Viperridae

SOURCE: IBGE Interactive Maps, 2005.

Table 8 - Endangered mammal families

Fauna	Family	Family
Mammals	Baleanidae	Cricetidae
	Balenopteridae	Dasypodidae
	Brandypodidae	Emballonuridae
	Callimiconidae	Felidae
	Callitrichidae	Mustelidae
	Canidae	Nymecophagidae
	Cebidae	Phyllostomidae
	Cervidae	Pontoporiidae
		Vespertilionidae

SOURCE: IBGE Interactive Maps, 2005

Table 9 - Endangered fish and insect families

Fauna	Family
Fish	Pimilodidae
	Coenagrionidae
	Lycaenidae

Insects	Nymphalidae Peripatidae Pieridae Pseudostigmatidae

SOURCE: IBGE Interactive Maps, 2005.

4.2.3. ANTHROPIC ENVIRONMENT

4.2.3.1. Land occupation

The first settlers in the region were European immigrants, mainly of Italian, German (Pomeranian) and Polish origin at the end of the 19th century, who founded a settlement on the banks of the Santa Maria River. Due to aspects of their culture, they named the village after São Roque, the patron saint of diseases. This settlement became a district under the name of São Roque, by state law no. 137 of 2 September 1982, subordinated to the municipality of Santa Teresa. It was elevated to the category of municipality after a popular plebiscite held on 25 June 1995, with the name of São Roque do Canaã, by state law no. 5147, of 18 December 1995, separated from the municipality of Santa Teresa (IBGE, 2011).

Historically, the municipality has been farmed by small and medium-sized rural producers, with successive fragmentations of rural properties into small family plots, as shown in Table 10 below, resulting in a disconnection of native forest cover and causing a major impact on gene flow between the different remaining fragments.

Table 10 - Municipal Land Stratification

Distribution	No. of rural properties	Percentage
Smallholdings - under 18.0 ha	528	48,84%
Small property* (from 18.0 to 72.0 ha)	495	45,8%
Medium-sized property* (Above 72.0 to 270.0 ha)	55	5,08%
Large property* (over 270.0	3	0,28%

ha)		
Total number of properties	1.081	100

* According to Normative Instruction No. 11 of 04 April 2003. São Roque do Canaã Fiscal Module = 18.0 ha; Minifundio = below 1 fiscal module; Pequena = from 1 to 4 fiscal modules; Média = above 4 up to 15 fiscal modules; grande = above 15 fiscal modules. Source: PROATER, 2008.

According to the Brazilian Institute of Geography and Statistics, the municipality has the following population characteristics (Table 11).

Table 11 - Population data for the municipality

Population Characteristics	
Population in 2010	.273 hab
Urban resident population	5,584 inhabitants
Urban resident population	5,689 inhabitants
Population density	32.92 inhab/Km^2

SOURCE: IBGE, 2011

It is worth mentioning that the municipality's rural population is distributed among small communities located mainly in the districts of Santa Júlia and São Jacinto. The municipality's Human Development Index in 2000 was 0.751 - considered average according to the United Nations Development Programme (UNDP), which aims to measure the living conditions of a population based on three dimensions: education, income and longevity (IBGE, 200).

The municipality has various social organisations such as: the Santa Luzia Rural Producers' Association, the Tancredinho Rural Producers' Association, the Sagrado Rural Producers' Association, the Santa Júlia Agrovillage Rural Producers' Association, the Saúde Residents' Association, the São Roque do Canaã Rural Workers' Union and Unicana. These work directly in the rural sector. The São Roque do Canaã Charitable and Cultural Association is also involved in recreation, culture and sport in the town centre.

4.2.3.2. Socio-economic profile

The municipality's main economic activity, with 52% of the economically active population, is in the agricultural sector, with beef cattle and coffee growing, fruit and vegetable production and sugar cane cultivation, the latter mainly based on smallholdings and family labour (PROATER, 2008).

Despite the prominence of coffee growing in the municipality's economy, it's important to note that most of the producers carry out some kind of complementary agricultural activity, generating alternative sources of labour and income. Thus, the cultivation of vegetables (especially tomatoes), sugar cane and fruit growing (guava, bananas, pine cones and mangoes) have emerged as secondary activities (PROATER, 2008).

In the industrial sector, the municipality of São Roque do Canaã is a regional highlight in the red ceramics industry, mainly in the production of roof tiles, wood frame factories and the production of "cachaça" brandy, the latter located mainly in the communities of São Dalmácio and São Sebastião. The local tradition of making this drink dates back to the beginning of the 20th century, when the first sugar cane mills were set up. Today, the region has 26 sugar cane mills producing sugar cane in the mould of a family-run agro-industry. The significant production and quality of the beverage produced has led to the creation of a cooperative of sugar cane producers called UNICANA (PROATER, 2008).

According to the Brazilian Institute of Geography and Statistics, in 2008 the municipality had a Gross Domestic Product (GDP) of R$ 88,483.911 thousand and a GDP per capita of R$ 8,203.59 (IBGE, 2010).

4.3. OVERVIEW OF THE CERAMICS INDUSTRY IN THE MUNICIPALITY

4.3.1. Business Profile

According to the North Central Region Pottery Industry Union (Sindiolaria - Norte), there are eight red ceramics industries in the municipality, characterised as small and medium-sized companies, five of which are located in the urban area and three in the rural area, both of which are of great economic and social importance to the municipality as a source of labour and revenue generation, responsible for a significant part of the municipal GDP, employing around 750 direct jobs and 1,200 indirect jobs, such as truck drivers, specialised services in machine and equipment maintenance, outsourced transport of employees, among others.

These enterprises basically produce ceramic blocks for structural sealing, with an estimated daily production of 100,000 units, and 300,000 roof tiles, both of which come in various models. However, with this volume of roof tile production, the municipality leads the way as the largest producer in the northern region of the state.

Products are marketed in two ways, the first being directly to the end consumer, in this case focussing on the local and corporate market (construction companies, which account for only 23% of the sector's turnover) and the government. On the other hand, indirect distribution takes place through wholesalers and specialised retailers (home centres and building materials shops), which are the main channels for getting products to end consumers in neighbouring states such as Minas Gerais, Bahia and Rio de Janeiro.

These industries basically use common clay as a raw material, with an estimated consumption of 1,800 tonnes per day, extracted from deposits located in the municipality and neighbouring municipalities (Santa Teresa and Colatina), which is due to the fact that almost all the deposits in the municipality are already registered or exploited. Energy sources include electricity, diesel oil, although the latter is not used by all industries (used in generators daily from 5.30pm to 8.30pm) at peak times, and wood (sawdust and reforested wood), the latter with a monthly consumption of 10,350 m .^{3}

According to the data obtained from the interviews with entrepreneurs, the sector is going through a labour shortage, which is leading to production restructuring, especially on a technological basis, through the current trend towards automation of some processes and procedures, which will certainly change the pattern of competition in the coming years. Another obstacle mentioned by businesspeople is the high tax burden and delays with public agencies in analysing licensing requests, followed by the high complexity of regulations, the cost of investments needed to meet requirements, and the high cost of preparing studies and projects to present to the environmental agency. This characterises the existence of bureaucratic, non-systemic and disjointed procedures for obtaining a licence.

Table 5 - Overview of the ceramics industry in São Roque do Canaã - ES

General Aspects of the Ceramics Industry	
Number of Production Units (Companies)	**8**
Number of Direct Jobs (ED):	**750**
Men	**600**
Women	**150**
Number of indirect jobs (IE)	**1.200**
Predominant level of education (ED)	**Basic level**
Average age of employees	**35 to 45 years old**
Salary range	**750 to 2,000 reais**
Pieces produced per day (roof tiles)	**300.000**
Pieces produced per day (blocks)	**100.000**
Consumer market (tiles and blocks)	**ES, MG, BA, RJ**
Clay consumption	**1,800 tonnes/day**
Firewood consumption	**10,350 m /month^3**

SOURCE: The author, 2011.

4.3.2 Overview of the production process

The production process in the ceramics industry is characterised by two distinct stages: the primary stage (which involves exploiting the raw material - in this case, clay) and the transformation stage (to produce the final product).

1ª stage - Select the right clay for the production of tiles or ceramic blocks. After selecting the clay extraction site, laboratory analyses and production tests must be carried out, involving the malleability of the clay, characteristics compatible with the geometry of the tiles to be produced, as well as evaluating the strength of the dry clay, the result of the tile evaluating bending and the level of porosity, among other items.

2ª stage - Seasoning: the clay extracted from the quarry is sent for a storage period of no less than six months. The main aim of this seasoning is to improve the plasticity of the clay to be used in the manufacture of roof tiles. The purpose of this period is also to wash out soluble salts, decompose organic matter and reduce the tensions caused by the breaking of chemical bonds.

After this seasonal period, the raw material is ready for the ceramic tile manufacturing process, which involves a four-stage process: preparing the mass, shaping the pieces, drying and firing. In order to obtain a quality product and minimise defects and losses, strict monitoring is required at each stage.

Preparing the dough

The ceramic clay for roof tiles is prepared in such a way as to obtain an ideal composition of plasticity and fusibility, to make it easier to work with and mechanically resistant to firing in a special kiln for this purpose. The clay is usually prepared by mixing a "fat" clay, which is characterised by high plasticity and fine granulometry, with a "lean" clay, which is less plastic and coarse-grained, rich in quartz, which acts as a plasticity reducer. After mixing, the mixture is moistened to an average content of 20% and homogenised to form the ceramic products.

The following equipment is normally used at this stage: mill, feed bin, crumbler or disintegrator, mixer and rolling mill.

Extrusion/Blackout

At this stage, a clay mass is sought that is a paste with good plasticity and adequate rigidity, which is forced through a mould to form a continuous column, which can be cut into appropriate lengths, according to the models that have been previously defined for production.

The usual extrusion machine is known as a maromba or extruder, and its function is to homogenise, break up and compact the ceramic masses to shape the desired product.

Pressing

The clay paste is pressed at this stage. The clay paste is pressed immediately after the extrusion stage, which is an intermediate stage in the forming process, followed by pressing after the extruded column has been cut. The method consists of placing the granulated mass with a lower moisture content in a rubber mould or other polymeric material, which must be closed hermetically, resulting in a compressed process in the previously defined format.

Drying

After forming, the drying stage begins. This process is a very important operation in the manufacture of structural ceramics, requiring special care to ensure that the water contained in the products is slowly and evenly eliminated throughout the ceramic mass to avoid possible defects in the piece such as cracks, warping or breakage.

The drying stage can be carried out in two ways:

Natural drying: in the open air, close to the ovens to take advantage of the circulating heat, for a period of 6 to 12 days, depending on the relative humidity of the air in the drying area. Artificial drying: drying takes place in static, continuous or semi-continuous dryers, with the controlled introduction of hot air from furnaces or ovens.

Burning

Firing is the most important stage of any production process. It is during this stage that the various properties of clays are manifested through the physical, chemical and mechanical transformations caused by the action of fire. The firing process involves four phases.

1. Heating or preheating: characterised by gradual heating to remove residual water, without causing defects in the ceramic piece caused by differential contractions during the expulsion of the remaining moisture, over a period of 8 to 13 hours, with the kiln temperature reaching 650°C;

2. Hard firing or firing: this process starts at around 650°C and can be increased at a greater rate up to 950°C or 1000°C. It is during this phase that the chemical reactions take place that give the ceramic body its characteristics of hardness, stability, resistance to various physical and chemical agents, as well as the desired colouring;

3. Plateau: the maximum firing temperature is maintained for a certain period of time, so that the chamber is as close as possible to the desired temperature for the entire length of the furnace;

4. Cooling: carried out gradually and carefully to avoid cracking, through the chimney or by using the heat for the dryers, over a period of around 38 to 50 hours.

4.3.3. **Flowchart of the production process**

SOURCE: The author, 2011.

CHAPTER 5

RESULTS AND DISCUSSION

5.1. ENVIRONMENTAL DIAGNOSIS OF CLAY EXTRACTION AREAS

The data collected in the study area revealed various irregularities and environmental impacts. Among these is the illegal exploitation of clay, which has caused damage to the physical, biotic and socio-cultural environment. In order for the ceramics industries to gain access to the clay extraction areas, it is necessary to open roads or improve access routes, as shown in figure 5, which has initially led to some environmental impacts, such as altering the natural landscape, generating noise, emitting particulate matter, emitting gases, the possibility of leaks and contamination by oils and greases, altering the local road system, the possibility of accidents to species of fauna due to being run over, noise pollution, stress on natural vegetation, among others.

Figure 5 - (A) Opening of roads to access the extraction area (B), damage to the local flora due to the deposition of particulate material from the transport of the raw material (clay).

SOURCE: The author (2011).

The next step in clay extraction must be to prepare the area, which consists of removing or "stripping" the vegetation from the top layer of soil, as seen in (Figure 6). This "stripping" removes the most fertile part of the soil for agriculture, rich in organic matter and nutrients available for plants. Also at this stage, the loss of biodiversity is one of the main ecological damages, where we see a reduction in the sustainability potential of the systems, jeopardising the existence of animal and plant species. As discussed by Sánchez (2008), soil removal processes involve the loss of the surface organic horizon, to which is added the loss of biodiversity of organisms associated with the soil microbiota and mesofauna.

Figure 6 - Preparing the area for clay extraction.

SOURCE: The author (2011).

The vast majority of clay extraction areas are located in the interior of the municipality and the most worrying thing is that many of them are located in Permanent Preservation Areas - PPAs, mainly in marginal areas of water bodies. However, there are also exploited areas in the urban area. In both areas, the irreversible modification of the landscape stands out.

Many of these explored or being explored areas located in the countryside belong to small landowners, who rent them out to ceramics companies to extract the clay. The amount paid to the owners is generally between R$65.00 (sixty-five reais) and R$80.00 (eighty reais) per 12 (twelve) cubic metre bucket.

In the APP areas, we observed that clay extraction has caused damage to the flora, fauna, soil and water environments, the latter because sediment is carried into the watercourses through soil erosion, causing silting, increasing turbidity, affecting different aquatic communities and the trophic chain, among other things. Another alarming fact is that, regardless of whether the extraction takes place in an irregular area, there are no mitigating control measures or sediment containment devices.

It is believed that all the changes resulting from the adoption of inadequate practices associated with the maintenance of these areas go beyond the boundaries of a rural unit, acquiring, as a whole, great social importance with impacts on the urban environment, affecting society as a whole. As discussed by Farias et. al. (2006), one of the emblematic examples of this is the question of the availability of water resources, where the frequent scarcity of water for supply in various urban centres, as well as the rationing of electricity supplies caused by the low level of reservoirs, could be attributed, in part, to the chronic degradation of riparian forests and spring areas in various Brazilian river basins over the last few decades.

Another aspect observed in many "deactivated" quarries and those being exploited was the overexploitation of these environments to remove the raw material (clay) until it reached the water table (figures 7 and 8), caused by the opening of pits varying in depth from 1 to 4 metres. According to local residents, this procedure has interfered with the flow of groundwater and surface water, as well as completely altering the natural dynamics of the area by flooding the pit, causing the destruction of habitats, displacement of fauna, risks to human health due to the possible proliferation of vectors, among other things. The excessive exploitation of the deposits was the reason for the intervention of the Public Prosecutor's Office, by means of a Conduct Adjustment Agreement (TAC) with ceramists in the region, who undertook to adjust the behaviour considered illegal.

Figure 7 - Groundwater outcrop during clay extraction

Source: The authors, 2011.

Figure 8 - Abandoned area, after clay extraction and outcropping of the water table.

SOURCE: The author (2011).

In the socio-economic sphere, clay extraction was observed in livestock areas and coffee production areas (Figure 9), which could lead to a reduction in agricultural areas, changing the way land is used and occupied by clay extraction activities. Clay extraction has compromised the productive capacity of soils, as well as causing economic losses by reducing areas for primary production and inducing the search for new agricultural frontiers, thus increasing the pressure to deforest new areas and land.

Figure 9 - (A) Clay extraction in livestock areas, (B) in coffee production areas, reducing agricultural areas.

SOURCE: The Author, (2011).

In the same context, irreversible environmental impacts on the local landscape have been observed in the socio-cultural sphere. Since the landscape reveals the meaning of a society's relationship with the environment, it is where history is recorded and totalled. Thus, once modified, the landscape can jeopardise spatial references to popular memory and culture.

In addition to the activities and impacts described above, another form of environmental damage observed is the irregular decommissioning of quarries (clay extraction areas), as shown in Figure 10. Most of these sites remain abandoned, without any human intervention to rehabilitate the area. The flora has been slowly restored over time.

Figure 10 - Former clay extraction areas without proper rehabilitation of the area.

SOURCE: The Author, (2011).

It can be seen in the field that there is no supervision or monitoring in a large part of the "exploited" deposits that have been deactivated, and there is no retaliation of the "pits" to prevent landslides and erosion, as shown in Figure 11 below.

Figure 11 - Collapse of old clay extraction pits without the pits having been properly filled in.

SOURCE: The author (2011).

In this regard, it should be noted that the environmental obligations governing the recovery of degraded areas after the extraction of clay or other minerals are dealt with by Decree No. 97.632 of 10 March 1989, as there is no specific infra-constitutional law. Therefore, the environmental rehabilitation and stabilisation of areas degraded after clay extraction is a legal obligation.

It is also important to mention that art. 55, sole paragraph of the Environmental Crimes Law No. 9.605, of 12 February 1998, defines as a crime and administrative infraction, subject to a fine, the fact of failing to recover the mined area under the terms of the determination of the competent environmental agency.

As described by various authors, in the case of pits resulting from clay extraction, a number of uses can be considered to fulfil this objective. These include their use for fish farming, plant production for agricultural, livestock or forestry purposes (Rodrigues, 2001; Pralon, 1999; Schiavo, 2001), or simply to restore the biological balance of the site area (Lourenzo, 1991). However, in any of these cases, it is necessary to verify the existence of minimum conditions for the implementation of the intended use.

It should also be noted that for rehabilitation to be successful in areas degraded by mining, it is important to establish a programme that must include the planning of the mining activity, from its conception phase to the phase following the end of mining, known as mine closure.

It should be emphasised that the peculiarity of the issue of closing a mine stems from the process of changing the use of the area, and it is essential to observe the legal impositions that derive from this fact, relating to the closure of the mine itself, the need for licensing of the new form of use, and the responsibility of the miner for complying with the obligation to carry out the recovery plan for the degraded area approved by the competent environmental agency (SOUZA, 2002).

Potters are obliged to implement a rehabilitation plan for areas degraded by mining activities, approved by the competent environmental agency, which includes the future use of the area of influence of the quarry after its closure. In the study area, some quarries were also being rehabilitated. However, the initiative taken by the companies reflects more of a response to a demand from the environmental agencies than taking a socially and environmentally responsible stance.

In loco, it can be seen that the planting of flora species for the rehabilitation of quarries has been taking place inappropriately, where the choice of species to be planted does not always take into account their ecological importance in the environments to be recovered, using exotic species as alternatives, without ecological adaptation and with a risk of becoming invasive species.

A floristic and phytosociological study of the region should be carried out to find out which native species could be introduced into the rehabilitation areas. Annex 2 contains a list of 97 native species of riparian forest in the region (which could be used for planting in the rehabilitation of the quarries). The species were identified by Brito (2006) apud PRAD (2008), through a floristic and phytosociological study carried out along the stretch between the source of the Santa Maria do Doce River and the municipality of São Roque do Canaã, as shown in figure 12 below.

It should be emphasised that planting should prioritise, as far as possible (depending on the availability of seedlings in local nurseries), the species with the highest importance value, which is obtained by adding up the relative values of abundance (density), dominance and frequency for each species.

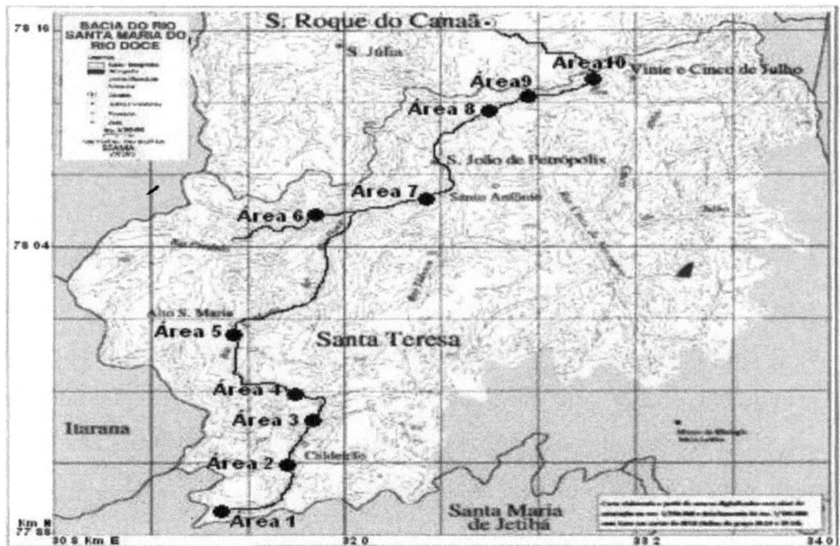

Figure 12 - Location of the phytosociological survey areas

SOURCE: SEAMA (2003), modified by BRITO (2006) apud PRAD (2008).

A list of some of the most important species for the region is shown in Table 12, which could also be used by ceramics companies when making decisions about the purchase of seedlings.

Table 12 - List of species with the highest Importance Value.

Selected species	Common name	GE	VI
Anadenanthera macrocarpa	Angico	Yes	21.42
Peschiera laeta	Dairy	Sc	19.27
Guapira opposita	Maria-mole	St	10.93
Machaerium hirtum	Barreiro	Pi	10.83
Euterpe edulis	Sweet palm	C	9.57
Neoraputia alba	Arapoca	C	8.34
Matayba arborescens	Turkey pitch	C	7.32
Croton floribundus	Capixingui	Pi	6.96
Alchornea triplinervia	Boleiro	Yes	6.51
Guarea grandiflora	Creep	C	5.68
Rollinia laurifolia	Araticum	C	5.62
Bauhinia forficata	Cow's foot	Pi	5.17
Ramisia brasiliensis	Ganansaia	Sc	4.99
Genipa americana	Jenipapo	St	4.94
Myrcia fallax	Lanceira	Pi	4.77
Hieronyma alchorneoides	Licurana	Yes	4.66
Feuilleea edulis	Ingá	C	3.71
Alseis floribunda	St Paul's quinoa	St	3.53

Legend: **GE** = ecological groups; **Pi** = Pioneer; **Si** = Early Secondary; **St** = Late Secondary; **C** = Climatic; **SC** = uncharacterised; **VI** = value of importance.
SOURCE: PRAD (2008).

Another technique that has recently been discussed in rehabilitation programmes for degraded areas, and which has boosted the recovery of these areas, is replacing the mere application of agronomic or silvicultural practices for planting perennial species with a concept of reconstructing the interactions of the ecological community, including recognising the importance of frugivore-plant interactions in this process and recommending the planting of native pioneer and early secondary species that are attractive to fauna.

The planting of pioneer and early secondary zoochorous species can therefore be used in the rehabilitation of degraded areas. Since frugivores attracted by the zoochoric species used in the planting not only disperse the seeds of these plants, but also bring with them seeds of other native species, increasing the specific richness of the area (PRAD, 2008).

5.2. TEXTURAL CLASSIFICATION OF SOIL FROM CLAY EXTRACTION SITES

Soil from 16 (sixteen) clay extraction quarries, chosen by lot and located in the rural region of the municipality, exploited by different ceramics industries, were submitted to textural classification. 8 of the extraction areas were used for the production of roof tiles and the other 8 were used for the production of ceramic structural sealing blocks. To carry out the laboratory analyses, a composite soil sample was collected from each deposit. The granulometric analyses of texture and the classification of the results were determined according to the standard methodology of EMBRAPA (1997). The results obtained are described in the following tables.

Table 13 - Textural classification of the samples (for roof tile production).

*Locations	Particle size (%)			Textural Classification
	Clay	Silt	Sand	
L 1.1	75,8	15,3	8,9	Very clayey
L 1.2	62,4	30,4	7,2	Very clayey
L 1.3	62,1	29,8	8,1	Very clayey
L 1.4	61,6	31,1	7,3	Very clayey
L 1.5	60,2	25,1	14,7	Very clayey
L 1.6	58,6	28,1	13,3	Clay
L 1.7	58,4	26,7	14,9	Clay
L 1.8	58,1	27,4	14,5	Clay

*Samples collected from quarries exploited for the production of roof tiles.

Table 14 - Textural classification of the samples (for the production of ceramic blocks).

**Locations -	Particle size (%)			Textural Classification
	Clay	Silt	Sand	
L 2.1	52,6	14,2	33,2	Clay
L 2.2	51,4	15,6	33,0	Clay
L 2.3	51,1	14,8	34,1	Clay
L 2.4	50,2	13,9	35,9	Clay
L 2.5	48,7	15,4	35,9	Clay

L 2.6	46,1	14,6	39,3	Clay
L 2.7	42,7	16,6	40,7	Clay
L 2.8	42,5	15,9	41,6	Clay

**Samples collected from quarries exploited for the production of ceramic blocks.

In view of the laboratory results obtained, it can be said, as was already known from what was described by representatives of the ceramics industry (based on experience), that the raw material (soil with a predominance of clay) is different for the production of roof tiles and for the production of ceramic blocks. On average, the raw material for roof tile production is more clayey (62.15%) and silty (23.73%) than for ceramic block production. The raw material used to make ceramic blocks is different in that it has a low silt content (15.12%) and a high sand content (36.71%).

Both for the production of roof tiles and ceramic blocks, we can see that this difference is directly related to the clay, silt and sand content of the soil, as seen in Tables 13, 14 and Figure 13 below.

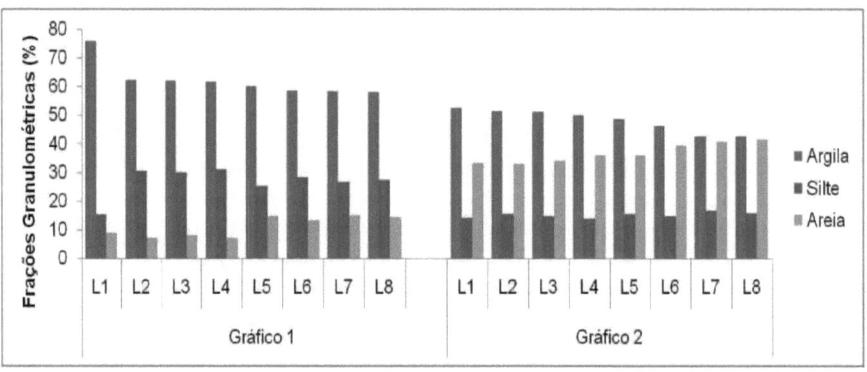

Figure 13 - Graph 1- Representation of the sand, silt and clay content of the quarries (used to make roof tiles) and graph 2 of the quarries (used to make ceramic blocks).

The results of the samples carried out showed that for the production of roof tiles by the ceramics industries in the municipality, soils with high clay and silt contents are used, with "very clayey soil" as the textural class. However, clay class soils are also used, but these are mixed with soils with a higher clay content, so it is expected that they will produce a material of good quality and mainly with low porosity.

The results of the samples analysed also show that soils with a lower clay content and a high sand content are used for the production of ceramic blocks, which comprise the textural class of "clay soils". These soils, which have a lower clay content and are abundant in the municipality, are used to produce ceramic blocks because the final product is not compromised.

5.2. TABLE OF THE MAIN ENVIRONMENTAL ASPECTS AND IMPACTS OF CLAY EXTRACTION

Through on-site data collection at the extraction sites, information obtained from representatives of the municipality's ceramics industries and bibliographical research, tables were drawn up identifying the main environmental aspects and impacts of clay extraction, as described in Table 6 below:

Table 6 - Main environmental aspects and impacts - implementation phase

Activity	Aspects	Impacts
Land acquisition (raw materials)	Income generation	Payment of royalties to owners and indemnities
Licensing of activity	Inventory of resources, delimitation of areas of occurrence, income generation	Hiring specialised labour for consultancy services
Collection of material (soil) for laboratory analyses	Planning	Hiring a specialised laboratory to carry out analyses
Drawing up the extraction plan	Planning, income generation	Hiring specialised labour for consultancy services
Opening or improving access roads	Alteration of the natural landscape, generation of noise, particulate matter, waste, gas emissions, oil and grease leaks,	Noise pollution, air pollution, neighbourhood disturbance, soil and water contamination by oils and greases, damage

	alteration of the local road system possibility of accidents	to local flora and fauna

SOURCE: Adapted from Bacci et. *al.,* (2006); Sánchez (2010).

Table 7 - Main environmental aspects and impacts - extraction phase

Activity	Aspects	Impacts
Preparing sites for the disposal of topsoil and tailings	Alteration of the natural landscape, generation of noise and particulate matter	Contamination of watercourses and soil, modification of the local landscape, damage to local flora and fauna, noise pollution
Removal of vegetation or stripping of the deposit	Alteration of the natural landscape, generation of noise, particulate matter, scaring away of flora	Modification of the natural landscape, destruction of native vegetation, damage to local fauna contamination of water courses and soil, noise pollution
Temporary disposal of topsoil	Alteration of the natural landscape, generation of noise and particulate matter	Modification of the natural landscape, damage to local flora and fauna, noise pollution
Exploitation of the deposit by digging and stockpiling clay	Alteration of the natural landscape, generation of noise and particulate matter, alteration of air quality	Use of natural resources, modification of the natural landscape, erosion, noise pollution, disturbance of the

		neighbourhood, reduction of the area for primary production, increased sediment load, silting and contamination of water bodies, lowering of the water table, etc.
		groundwater table, alteration of groundwater flow, loss of spatial references to memory and popular culture, payment of royalties to landowners and indemnities
Movement of machinery and transport of raw materials	Generation of noise, particulate matter, waste, gas emissions, oil and grease leaks, alteration of the local road system possibility of accidents	Noise pollution, atmospheric pollution, neighbourhood disturbance, contamination of soil and water by oils and greases, damage to flora, scaring away and loss of fauna specimens by being run over, stress on natural vegetation
Stockpiling raw materials in the development's	Generation of sediment, particulate matter, change in air quality,	Air pollution, visual pollution, watercourse pollution and soil

courtyard	modification of the local landscape	contamination, increased sediment load, siltation
Processing	Generation of noise, particulate matter, waste, gas emissions, possibility of accidents, electricity consumption, wood consumption, water consumption, effluent generation, increase in tax collection	Tax collection, hiring local labour, noise pollution, use of natural resources, disturbance to the neighbourhood, occupational exposure of workers

SOURCE: Adapted from Bacci et. al., (2006); Sánchez (2010).

Table 8 - Main environmental aspects and impacts - decommissioning phase[1]

Activity	Aspects	Impacts
[1] **Stoppage of activities without specific planning for the rehabilitation of impacted areas**	Physical-Biotic Environment: Alteration of the groundwater and surface water flow regime, destruction of habitats, displacement of fauna, creation of new environments, Initiation or acceleration of erosion processes, instability of marginal slopes Socio-economic: Risk to human health due to the	Environmental pollution, Visual pollution, Pollution of water bodies Compromise of human health, Compromise of income for site owners

	possible proliferation of vectors caused by the accumulation of water in extraction "holes", replacement of economic activities, modification of land use, reduction in primary production, alteration or destruction of sites of cultural or tourist interest, reduction in landowners' income due to non-payment of royalties and compensation, reduction in the local road infrastructure.	

SOURCE: Adapted from Bacci et. al., (2006); Sánchez (2010).

Table 9 - Main environmental aspects and impacts - decommissioning phase[2]

Deactivation phase	Aspects	Impacts
[2] **Paralysis of activities with specific planning for the rehabilitation of impacted areas**	In order to compensate for and mitigate this phase of activity, the project must proceed in accordance with the conditions established by the licences issued by the competent environmental agency. Some measures are	Hiring services and specialised labour for consultancy services

	described below:
	Filling in the excavations in order to provide a suitable environment for the development of native vegetation and the planting and revegetation of specimens of the local flora through a phytosociological study of the region.

SOURCE: Adapted from Bacci et. al., (2006); Sánchez (2010).

5.3. INTERACTION MATRIX OF THE ENVIRONMENTAL IMPACTS OF CLAY EXTRACTION

By collecting primary data through on-site visits to the quarries, information obtained from the municipality's ceramics industries and bibliographical research, a matrix was drawn up for three different phases of clay extraction, as shown below. After selecting the activities and components of the environmental impacts, the matrix was filled in to identify the positive (+) or negative (-) impacts and their duration, whether temporary (T) or permanent (P), considering all possible interactions with the physical, biological and anthropic environment.

Table 15 - Environmental impact interaction matrix - implementation phase

	Activity	Impacts	Type	Duration of impacts on the environment		
				Physical	Biological	Anthropic
Impl	Land	Payment of royalties to	+	T	T	T

ementation phases	acquisition (raw materials)	owners and indemnities				
	Licensing the activity	Hiring specialised labour for consultancy services	+	T	T	T
	Collection and processing of material for laboratory analyses	Hiring a specialised laboratory to carry out analyses	+	T	T	T
	Drawing up the extraction plan	Hiring specialised labour for consultancy services	+	T	T	T
	Opening or improving access roads	Noise pollution, air pollution, neighbourhood disturbance, soil and water contamination by oils and greases, damage to local flora and fauna	-	P	P	P

Legend: (+) positive, (-) negative; P - permanent, T - temporary.

Physical Environment - studies climatology, air quality, noise, geology, geomorphology, water resources (hydrology, surface hydrology, physical oceanography, water quality, water use), and soil;

Biological Environment - studies the terrestrial ecosystem, the aquatic ecosystem and the transitional ecosystem;

Anthropic Environment - studies population dynamics, land use and occupation, standard of living, production and service structure and social organisation.

Table 16 - Environmental impact interaction matrix - extraction phase

	Activity	Impacts	Type	Duration of impacts on the environment		
				Physical	Biological	Anthropic
Extraction Phases	Preparing sites for the disposal of topsoil and tailings	Contamination of watercourses and soil, modification of the local landscape, damage to local flora and fauna, noise pollution	-	T	T	T
	Removal of vegetation or stripping of the deposit	Modification of the natural landscape, destruction of native vegetation, damage to local fauna contamination of watercourses and soil, noise pollution	-	P	P	P
	Temporary disposal of topsoil	Modification of the natural landscape,	-	T	T	T

	damage to local flora and fauna, noise pollution				
Exploitation of the deposit by digging and stockpiling clay	Use of natural resources, modification of the natural landscape, erosion, noise pollution, disturbance of the neighbourhood, reduction of area for primary production, increase in sediment load, silting up and contamination of water bodies, lowering of the water table, alteration of groundwater flow, loss of spatial	-	P	P	P

	references to memory and popular culture, payment of royalties to owners and compensation				
Movement of machinery and transport of raw materials	Noise pollution, air pollution, neighbourhood disturbance, soil and water contamination by oils and greases, damage to flora, scaring and loss of fauna specimens by trampling, stress on natural vegetation	-	T	T	T
Stocking raw materials in the development's courtyard	Air pollution, visual pollution, pollution of watercourses, increased sediment load,	-	P	P	P

Processing	siltation Tax collection, hiring local labour, noise pollution, use of natural resources, disturbance to the neighbourhood, occupational exposure of workers	+ -	P	P	P

Legend: (+) positive, (-) negative; P - permanent, T - temporary.

Table 17 - Environmental impact interaction matrix - decommissioning phase[1]

	Activity	Impacts	Type	Duration of impacts on the environment		
				Physical	Biological	Anthropic
Deactivation phases[1]	[1] Stoppage of activities without specific planning, for the rehabilitation of impacted areas	Environmental pollution, Visual pollution, Pollution of water bodies Compromise of human health, Compromise of income for site owners	-	P	P	P

Legend: (+) positive, (-) negative; P - permanent, T - temporary.

Table 18 - Environmental impact interaction matrix - decommissioning phase[2]

	Activity	Impacts	Type	Duration of impacts on the environment		
				Physical	Biological	Anthropic
Phases - Deactivation[2]	[2] Paralysis of activities with specific planning for the rehabilitation of impacted areas	Hiring services and specialised labour for consultancy services	+ -	P	T	P

Legend: (+) positive, (-) negative; P - permanent, T - temporary.

CHAPTER 6

CONCLUSIONS

Considering the current situation of most of the areas exploited for clay extraction by the ceramics industry in the municipality of São Roque do Canaã, it can be seen that the damage to environmental integrity is considerable. Although clay extraction is necessary, it is essential that there is a sustainable way of extracting it, so that environmental legislation is complied with and impacts can be minimised, especially those on the landscape.

The clay extraction sites reveal alterations and degradation of the environment, which often cause irreversible impacts on the flora, fauna, soil and water environments, since most clay extraction takes place in permanent preservation areas, especially on plains on the banks of watercourses. Another point worth highlighting is the abandonment of extraction areas without proper rehabilitation, which can cause irreversible impacts on the local landscape.

It is also worth mentioning that when companies restore flora, they introduce exotic species as alternatives, without ecological adaptation and with the risk of becoming invasive species. This attitude needs to change, as companies are in a position to carry out quality actions to rehabilitate areas, demonstrating social and environmental responsibility. However, a future with less impactful development will only be possible if companies adopt environmental management systems.

With regard to the textural classification of the soils in the quarries, they have distinctly different textures, with the soils used to make roof tiles being more clayey and silty, classified as very clayey, and those used to make ceramic blocks for structural sealing being less clayey and more sandy, classified as clayey.

Finally, it is important to stress that the local community should be aware of the company's environmental policy, philosophy, guidelines, targets and impact minimisation programme. In this way, the community has elements to assess the real and potential compromise to its quality of life. Communication, on the other hand, makes the company aware of the community's wishes and level of support. A good relationship between the company and the community not only brings benefits to the latter, but also reduces the company's environmental vulnerability and increases security in the viability of the enterprise.

It is hoped that this study will play a role in raising important issues for the region. Mainly because this is an area that has historically exploited natural resources extensively, without adopting a clear policy for managing these resources.

CHAPTER 7

BIBLIOGRAPHICAL REFERENCES

BACCI, D. L. C.; LANDIM, P. M. B.; ESTON, S. M. **Aspectos e impactos ambientais de quarreira em área urbana**. REM: R. Esc. Minas, Ouro Preto, 59(1): 47-54, Jan. Mar. 2006.

BERTONI, J.; NETO, L. F. **Soil Conservation**. São Paulo: Ícone, 2008. - 6th edition, 355p.

BRAZIL. **Brazilian Forest Code**. Law No. 12.727, of 17 October 2012.

BRAZIL. **Decree No. 97.632 of 1 April 1989**.

BRAZIL. **Environmental Crimes Law**. Art. 55 of Law No. 9.605, of 12 February 1998.

DIAS, R. **Gestão Ambiental: Responsabilidade social e sustentabilidade**. 1. ed. - 7 reprint - São Paulo: Atlas, p. 196, 2010.

DNPM - National Department of Mineral Production. Available at: << http://www.dnpm.gov.br/ >>. Accessed on 14 October 2011.

EMBRAPA - **Brazilian Agricultural Research Corporation**. Permanent Preservation Areas and Sustainable Development, 2003. Available at << http://cediapgeo.ourinhos.unesp.br/material/esenvolvimento_sustentavel_-_embrapa.pdf >>. Accessed on 20 August 2011.

EMBRAPA - **Brazilian Agricultural Research Corporation**. National Soil Research Centre (Rio de Janeiro, RJ). Manual of Soil Analysis Methods / 2nd Ed. Ver. Atual. - Rio de Janeiro, 1997.212p. : il.

EMCAPA/NETUP - **Empresa Capixaba de Pesquisa Agropecuária**. Natural zones of Espírito Santo, 1999. Available at: << http://hidrometeorologia.incaper.es.gov.br/caracterizacao/saoroquedoc. php >>. Accessed on 5th August 2011.

FARIA, S. M. de; LIMA, H. C. de; RIBEIRO, R. D.; CASTILHO, A. F.; HENRIQUES, J. C. **Nodulation in leguminous species from the Trombetas region, Oriximiná, state of Pará and their potential use in reforestation of bauxite tailings ponds.** Soropédica: Embrapa Agrobiologia, 2006. 24 p. (Embrapa Agrobiologia Documentos, 209).

FEEMA - State Foundation for Environmental Engineering. State Legislation. 1997.

Available at: << www.feema.rj.gov.br >>. Accessed on 18 August 2011.

FEITOZA, L. R. **Agroclimatic Chart of Espírito Santo**. EMCAPA, 1986.

FERREIRA, M. M.; JÚNIOR. M. S. D. **Soil physics**. Lavras: UFLA/FAEPE, **2001. 117 p. : Il. Lato Sensu (Specialisation)** Distance **Learning Postgraduate Course**: Soils and the Environment.

HEUSER, Cristiane. **Identification of Environmental Aspects and Impacts in a Small Metalworking Company**. Undergraduate Monograph. Joinville/SC - 2007.

IBAMA - Brazilian Institute for the Environment and Renewable Natural Resources. Available at <http://www.ibama.gov.br/leiambiental/home.htm#crimesamb>. Accessed 21 Aug 2011.

IBAMA. Brazilian Institute for the Environment and Renewable Natural Resources. **Manual for Recovering Areas Degraded by Extraction:** Revegetation Techniques**.** Brasília: IBAMA. 96p. 1990.

IBGE - Brazilian Institute of Geography and Statistics. Available at: << http://www.ibge.gov.br/cidadesat/topwindow.htm?1 >>. Accessed on 21 August 2011.

IBGE - Brazilian Institute of Geography and Statistics. Available at: <<http://biblioteca.ibge.gov.br/visualizacao/dtbs/espiritosanto/saoroquedocan aa.pdf>>. Accessed on 21 August 2011.

IBGE - Brazilian Institute of Geography and Statistics. Available at: << http://www.ibge.gov.br/cidadesat/topwindow.htm?1 >>. Accessed on 21 August 2011.

IPES - Jonas dos Santos Neves Institute. Planned Investments for Espírito Santo (2000 data). Available at:< http://www.ijsn.es.gov.br/print.asp?p=true&urlframe=contasregionais/divulgac aom.htm& > Accessed on: 22/06/2011

KLEIN, V. A. **Soil physics -** Passo Fundo: Ed. Universidade de Passo Fundo, 2008. 212 p. : ill.; 23 cm.

LOURENZO, J. S. **Natural regeneration of an area mined for bauxite in Poços de Caldas, Minas Gerais**. Master's dissertation. Viçosa, Federal University of Viçosa. 151p. (1991).

PARH - **Water Resources Action Plan for the Santa Maria Do Doce Analysis Unit** - PARH Santa Maria Do Doce. Ecoplan-Lume Consortium. June 2010.

PAVAN, F. L. F. R. **Analysing the Production Agglomeration of the Ceramics Sector in the State of Espírito Santo** / Francisco Luiz Feu Rosa Pavan. - 2009.

PERRONE; A.; MOREIRA, T.H.L. **História e Geografia do Espírito Santo**. 6. ed. Vitória: UFES, 2005.

PHILIPPI JR, A.; ROMÉRO, M. A.; DRUNA, G. C.; editors. - **Environmental Management Course**. Barueri, SP: Manole, 2004p. 1045 (Environmental collection 1). PRALON, A. Z. **Production of Mimosa caesalpiniaefolia seedlings, inoculated with arbuscular mycorrhizal fungi and rhizobium, in clay extraction waste mixed with ferkal waste**. Master's thesis. Campos dos Goytacazes, Universidade Estadual do Norte Fluminense. 70p. (1999).

PROATER. **Technical Assistance and Rural Extension Programme**. São Roque do Canaã City Council, ES. p.84, 2008.

PRAD - Degraded Area Recovery Project. Recovery of a stretch of riparian forest along the Santa Maria do Doce River. Escola Agrotécnica Federal de Santa Teresa-ES. p. 158, 2008.

Available from RedeAmbiente<< http://www.redeambiente.org.br/dicionario.asp?letra=D&id_word=262 >>. Accessed 18 Aug 2011.

RODRIGUES, L. A. **Growth and nutrient absorption by Eucalyptus grandis plants and legumes in response to inoculation with arbuscular mycorrhizal fungi and rhizobium**. PhD thesis. Campos dos Goytacazes, Universidade Estadual do Norte Fluminense. 101p. (2001).

SÁNCHES, L.E. **Environmental Impact Assessment: Concepts and Methods**. São Paulo: Oficina de textos, p. 495, 2010.

SCHIAVO, J.A. **Production of guava (Psidium guajava L.) and Acacia mangium Willd seedlings colonised with arbuscular mycorrhizal fungi in pressed blocks made from agro-industrial waste**. Master's thesis. Campos dos Goytacazes, Universidade Estadual do Norte Fluminense. 86p. (2001).

SEBRAE/ES - Brazilian Micro and Small Business Support Service. **Sector diagnosis of the red ceramics and pottery industry in Espírito Santo** (Final Report), 2009. 52 p.

SEBRAE/ES - Brazilian Micro and Small Business Support Service. **Sector diagnosis of the red ceramics and pottery industry in Espírito Santo, 2009**.

SEBRAE - Brazilian micro and small business support service. **Sebrae/Espm market studies,** 2008. Red ceramics. Available at <<

http://www.biblioteca.sebrae.com.br/bds/BDS.nsf/E1EC5/$File8DA6.pdf. >> Accessed 16 Oct. 2011.

SINDIOLARIA - NORTE: **Pottery Industry Union of the Centre North Region of the State of Espírito Santo**. Available at: << http://www.sindiolarianorte.com.br/associados.php >>. Accessed on 21 August 2011.

SINDIOLARIA - SUL: **Pottery Industry Union of the Southern Region of the State of Espírito Santo**. Available at :<< http://www.sindiolariasul.com.br/ >>. Accessed 21 August 2011.

SOUZA, M. G. **Mine Closure: Legal Aspects**. Article published in IBRAM magazine, (2002). Available at << http://www.brasilminingsite.com.br/artigos/artigo.php?cod=31&typ=1 >>. Accessed on 19 August 2011.

VALICHESKI, R.R.; MARCIANO, C.R.; POCIANO, N.J. **Economic evaluation of the reuse of areas degraded by clay extraction in Campos dos Goytacazes - RJ**. Revista Ceres - 56(1): 001-008, 2009.

VALVERDE, R. S.; **Elements of Corporate Environmental Management.** - Viçosa: Ed. UFV, 127p. 2005.

Printed by Books on Demand GmbH, Norderstedt / Germany